Alaska's
Wild Berries

and berry-like fruit

Verna E. Pratt

Alaskakrafts, Inc.
P.O. Box 210087
Anchorage, AK 99521-0087

Seventh Printing—2007

Library of Congress Catalog Card Number.......95-94273

ISBN 0-9623192-4-4

Printed in Korea by Samhwa Printing Co., Ltd.

All photographs are by Verna and Frank Pratt, unless otherwise credited.

Illustrations by Verna E. Pratt, Technical Editor—Verna E. Pratt, General Editor—Frank G. Pratt

Front Cover Photo by Heidi E. Clifford
"Denali Park in the fall", Bog Blueberry and Alpine Bearberry foliage.

Table of Contents

ACKNOWLEDGEMENTS

We wish to thank everyone who encouraged us to produce this book. A special Thank You to Carol Griswold who assisted with proofreading and to the following who contributed some valuable photographs that helped make it possible:

Clayton Antieau, Forrest Baldwin, Heidi Clifford, Dave DeMoss, Sally Karabelnikoff, Rhoda M. Love, Ruth Munson, Frank Lang, and Michael MacDonald.

DISCLAIMER

Gathering berries for food is an enjoyable pastime, but requires careful attention to detail. Positive identification in the field is the reader's responsibility. Neither the author nor publisher assumes any responsibility for adverse reactions due to individual sensitivities, allergies or faulty plant identification.

ABOUT THE AUTHOR

Much of her work has been in the field, leading botanical field trips to various parts of Alaska. Her knowledge of plant ecosystems qualifies her as a superior field botanist and one of the most highly respected authorities of Alaskan wildflowers.

She is widely known for her pressed Alaskan Wildflower Craft items and her classes presented through the Anchorage Community Schools Program. She has also lectured at various locations within the State of Alaska and; more recently taught classes as an adjunct lecturer at the University of Alaska, Anchorage and through the Alaska Wilderness Studies program.

In 1982, she founded the Alaska Native Plant Society, and served as its State President through 1988. She is a member of the Alaska Orchid Society (AOS), Alaska Native Plant Society (ANPS), Alaska Society of Outdoor and Nature Photographers (ASONP), American Rock Garden Society (ARGS), American Society of Media Photographers (ASMP), Arts and Crafts League of Alaska, New England Wildflower Society (NEWS), United Alaskan Artists (UAA), and the Wildflower Garden Club.

She has also authored *Field Guide to Alaskan Wildflowers, Wildflowers along the Alaska Highway,* and *Wildflowers of Denali National Park.* See order form in back of book.

INTRODUCTION

This book was written to satisfy a strong interest in edible wild berries. Recognizing and finding favorite or new berries is at the top of the list for homemakers interested in reducing food costs for their families. Berry picking excursions bring families out in droves during late summer, and even avid hikers can seldom resist the desire to bend down and sample the more tasty varieties.

If used properly, this book will help you to recognize and enjoy the wild fruits and berries of Alaska.

Some confusion may arise from the fact that horticultural species are now appearing in our natural landscapes, the seeds having been carried there by birds. Amongst these are European Mountain Ash, Choke Cherries, Cotoneaster and Strawberry.

Many areas near cities are over-picked and trampled, which hampers production for future years. Many parks in crowded areas of the "Lower 48" do not allow berry picking for just that reason. At the time of this printing, picking berries on parklands in Alaska was permissible.

Consistently finding a good berry patch can sometimes be a challenge, as many factors affect berry production. Warm sunny days are needed in the spring and early summer to lure pollinators out of hiding. Abundance of the proper pollinators at the correct time is most important, and habitat may determine what the pollinator is.

For instance; wetland plants are often pollinated by mosquitoes or gnats, low plants in dry areas by crawling insects, white flowers by moths, sweet smelling ones by bees or butterflies. Some species of butterflies may also be attracted to a particular flower color. Once the pollinator has done its part, it is important to have the proper balance of sunshine and moisture to produce plump, juicy berries. The lack of, or overabundance of, either one could produce poor quality berries or none at all. I encourage you to search for your own special berry patches. Learn to recognize these species when they are in bloom and return when the berries are ripe (most berries require about 8 weeks). Take time to explore and you will surely enjoy nature and this book to its fullest extent.

This guide is not a book for experts; so, to keep the wording simple, we have used the term "berry" (although not botanically correct in all cases) for most of the fruits mentioned. It is compact in size and can easily be used in the field. Complete sentences are avoided in most cases, and simple terminology is used to simplify the text. The glossary at the back of the book will explain the more complicated terms. Botanical names are also included since, in some instances, duplicate common names for different species can be very confusing. Small maps are on each page, and we have shaded in the areas where the berries are found.

Edible berries are listed first (with a smiley face, or if not flavorful with a neutral face), and pages are color-coded according to berry color for ease in finding the descriptions. Inedible and poisonous berries follow with a sad face. I can't caution you enough to take time to properly identify anything that you eat. There are plenty of safe berries in Alaska. When in doubt, don't eat the berries until you verify their edibility. You can contact a knowledgeable person or the Cooperative Extension Service for more assistance. Try to avoid roadsides as the berries may be contaminated by automobile exhaust fumes or other materials. Remember, don't pick all of the berries in one location. They are an important food source for birds and bears. Damage as little of the plant as possible and leave no litter behind.

Enjoy, and Happy Berry Picking!   V.E.P.

Skunk Currant (see page 24)

Photo—Forrest Baldwin

RED OR ORANGE BERRIES

Included in this section are edible berries that are usually shades of red or orange (sometimes yellow) when ripe. There are many berries that are red, therefore, causing much confusion. None, however are difficult to recognize once you look closely and consider all the factors. The reason that many berries are red is so they can be seen easily by birds and animals who eat them and carry their seeds to a new location. Red is also less visible to insects, who might nibble away at the berries and destroy their possibility of reproduction.

Probably the most widely used berries are Cranberries. Bog Cranberry which bears single berries on weak stems has excellent flavor, but is difficult to pick. Therefore, Low-bush Cranberry, which grows in bunches and is plentiful is much more widely used. This is not a true Cranberry, but the flavor and texture, allow it to be used in much the same way as true Cranberries. There are many uses for this berry.

Raspberries are probably the 2nd most widely used though often they are only used for making jelly, which is probably the primary use of wild berries by most people. Do not overlook sour berries as they usually make excellent jelly or syrup. Mild flavors can be enhanced by the addition of some sour berries.

Estimating the amount of pectin when mixing berries is difficult if a recipe is not available. Try my freezer test method. Quickly chill a small amount of jelly in your freezer. If, after it has cooled, it hasn't jelled properly, add a bit more liquid pectin to the boiling jelly and test again. Of course, runny jelly can also be used for pancake syrups saving you the extra effort of adding more pectin.

RED BEARBERRY
Arctostaphylos rubra

Habitat: Bogs, tundra, moist woods below tree line.

Form: Sub-shrub — 2 to 5" (5 to 12 cm), mat forming.

Leaves: Deciduous — from base, spatulate, with wavy margins and course veins, yellow to red in fall.

Flowers: May to early June — greenish white, urn shaped, 5 joined petals, flowering as leaves are opening.

Fruit: August — berry, red, round, soft, juicy, translucent, under leaves.

Uses: Raw, cooked — used mostly as an extender in poor berry production years.

Red Bearberry

BOG CRANBERRY
Oxycoccus microcarpus

Habitat: Wet mossy bogs, wet tundra.

Form: Trailing shrub.

Leaves: Evergreen — alternate, ovate, small 1/6" (4 to 5 mm), dull, dark green above and lighter beneath.

Flowers: June — light pink, 5 reflexed petals, very small on a curved stem.

Fruit: September — berry, maroon, oval, small firm, tart, tasty, opaque, one per stem, lying on moss. Best after frost.

Uses: Raw (freeze well), cooked — sauce, jelly, jam, juice, relish, bread, cookies, pie (excellent flavor).

Other species: *Oxycoccus palustris* — similar variety, larger flowers and berries, extreme southern portion of Southeast Alaska.

Photo— Forrest Baldwin

Bog Cranberry

LOW-BUSH CRANBERRY
MOUNTAIN CRANBERRY
LINGONBERRY
PARTRIDGE BERRY
Vaccinium vitis-idaea

Habitat: Hummocks in bogs, woods, dry tundra.

Form: Low upright shrub — 3 to 8" (8 to 20cm).

Leaves: Evergreen — alternate, hard, shiny, oval, with edges rolled under.

Flowers: June — pinkish-white, bell-shaped, 5 joined petals — terminal cluster.

Fruit: September — berry, maroon, round, firm, in clusters, small, opaque, tart, tasty, mealy.

Uses: Raw (freeze well), cooked — sauce, jelly, jam, relish, juice, bread, cookies, pie, tea, liqueur (excellent flavor).

Comments: Produces best berries in open woodlands and old burned or cleared forests.

Low-Bush Cranberry

15

KINNIKINNICK
BEARBERRY
MEAL BERRY
Arctostaphylos uva-ursi

Habitat: Dry woods, exposed sites.

Form: Sprawling shrub — tap root, long decumbent branches.

Leaves: Evergreen — alternate, leathery, smooth above, rough and dull below, spatulate.

Flowers: Late May to early June — pink and white, urn-shaped, 5 joined petals, clusters near ends of branches.

Fruit: Late July through winter — berry, round, opaque, soft, mealy, orange to red.

Uses: Not generally used (dry, little flavor).

Kinnikinnick

TIMBERBERRY
NORTHERN COMMANDRA
PUMPKIN BERRY
DOGBERRY
Geocaulon lividum

Habitat: Dry woodlands

Form: Perennial plant — up to 8" (20cm), with creeping rootstock.

Leaves: Deciduous — alternate, long, oval, green to brown, sometimes mottled or variegated due to it's somewhat parasitic characteristics of feeding on roots of other plants.

Flowers: May to early June — green, small, cup-shaped, 5 sepals, close to main stem.

Fruit: July to August — berry-like, orange, soft.

Uses: Generally not used, not tasty.

Timberberry

BUNCH BERRY
GROUND DOGWOOD
CANADIAN DOGWOOD
Cornus canadensis

Habitat: Woodlands and low alpine slopes.

Form: Perennial, up to 5" (12cm), from creeping rootstock, with 2 small bracts below leaves.

Leaves: Deciduous — some retained 2 years, ovate, 4 to 5 in a whorl just below the flower, arcuate veins.

Flowers: June — greenish, 4 sepals, in a dense cluster in the center of 4 white pointed bracts.

Fruit: August & September — berry (drupe), orange to red, soft, in cluster above leaves, white pulp, sweet, insipid.

Uses: Raw or cooked — not generally used, not very tasty, edibility somewhat questioned by some authors.

Similar species: Swedish Dwarf Cornel, Lapland Cornel, *Cornus suecica*, has leaves in pairs on stem, flowers are wine to maroon color, found mostly on alpine slopes and tundra.

Comments: Frequent hybridization occurs between these species where their ranges overlap. (See small map)

Canadian Dogwood

Bunchberry

Lapland Dwarf Cornel

21

RED HUCKLEBERRY
Vaccinium parvifolium

Habitat: Moist woods.

Form: Shrub — upright up to 8' (2.5m). Twigs thin, greenish and angled, often with ridges.

Leaves: Deciduous — light green, thin, oval, slightly toothed when young, up to 1-1/4" (32cm) long.

Flowers: April to May — green or yellow to pinkish, urn-shaped.

Fruit: August — berry, red, soft, juicy, glossy looking.

Uses: Raw, cooked — jelly, jam, pie, syrup (Excellent flavor).

Photo— Forrest Baldwin

Red Huckleberry

Photo—Ruth Munson

RED CURRANT
Ribes triste

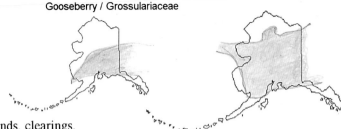

Habitat: Moist woodlands, clearings.

Form: Shrub — upright, 2 to 3 feet (up to 1 meter), alternate branches, shredding bark.

Leaves: Deciduous — alternate, toothed, 5 lobed (like maple leaves), 3 to 5 main veins, small and large leaves have same shape. Red in fall, winter buds brown, clustered at ends of branches. This is the only currant in Alaska that does not have strong skunk-like odor.

Flowers: May to early June — rosy-red, shallow cup-shaped, 5 tiny petals, 5 rounded sepals, hanging in long chains.

Fruit: July to early August — berry, red, soft, transparent, sour, seedy.

Uses: Raw, cooked, dried — Jelly, juice, liqueur (excellent flavor), strain out seeds,

Similar species: Skunk Currant or Fetid Currant (*Ribes glandulosum*) – sprawling shrub up to 30" (75cm), in moist woodlands. (See small map) Leaves are strongly toothed and have a strong odor. Berries red with glandular tipped hairs. See page 8.

Red Currant

HIGH-BUSH CRANBERRY
SQUASH BERRY
MOOSE BERRY
Viburnum edule

Habitat: Woods and meadows from lowlands into alpine.

Form: Shrub — upright, up to 8' (2.5m), opposite smooth branches.

Leaves: Deciduous — opposite, varied shapes, lobed and maple-like on lower branches, upper narrow toothed, coarse-veined (3 main veins). Buds red in winter.

Flowers: June — white to pinkish, 5 petals joined at base, in upright clusters.

Fruit: August — berry, red or orange (occasionally yellow), very sour, juicy, translucent, large flat seed, berry cluster may droop from weight.

Uses: Raw, cooked — jelly, jam, catsup, syrup, fish bait. Best cooked, seeds need to be removed. (Good flavor).

High-Bush Cranberry

WATERMELON BERRY
TWISTED STALK
WILD CUCUMBER
Streptopus amplexifolius

Habitat: Moist woods and meadows from sea level to low alpine.

Form: Perennial plant — 18" to 40" (up to 1 meter, sometimes taller), branched, arching stems, lower stem often thick and has some dark hairs, especially in spring.

Leaves: Deciduous — alternate, lance-shaped, sharply pointed, smooth edges, parallel veins, clasping the somewhat "zig-zag" stem. (In some areas, you might find varieties that have leaves with wavy edges).

Flowers: June — white to cream, 6 reflexed tepals, single on a twisted stem, under a leaf.

Fruit: August to September — berry, red to maroon, oval, juicy, sweet, mild flavor and seedy.

Uses: Raw, cooked — jelly, syrup, extender for other berries.

Watermelon Berry

ROSY TWISTED STALK
PINK-FLOWERED WATERMELON BERRY
Streptopus roseus

Habitat: Moist woods.

Form: Perennial plant — 18' to 36" (up to about 1 meter), slightly arching, unbranched, slender, hairless.

Leaves: Deciduous — alternate, clasping the stem, lance-shaped, slightly hairy edges and parallel veins.

Flowers: June — bell-shaped, pink, 6 joined tepals.

Fruit: August and September — red, round, soft, shiny.

Uses: Raw, cooked — jelly, syrup, extender for other berries.

Similar Species: *Streptopus streptopoides* is found in moist places at low elevation, mostly in Southeast Alaska. Plant is unbranched and hairless. Flowers are on a curved stem. Tepals are separate.

Rosy Twisted Stalk

31

SOAPBERRY
SOOPALALLIE
BUFFALO BERRY
Shepherdia canadensis

Habitat: Dry areas, woodlands, gravel bars.

Form: Shrub — up to 4' (1.25m), opposite branches, brownish scales (like sandpaper), young twigs coppery.

Leaves: Deciduous — opposite, oval, dark green above, lighter below with rusty scales. New winter buds coppery.

Flowers: May — yellow, 4 sepals, very small, male and female on separate shrubs.

Fruit: August — berry, red, oval, with some rusty scales, translucent, on main stem, bitter.

Uses: Not generally used now, but was traditionally used as a dessert topping by adding sugar to the whipped froth.

Male flower

Soapberry

SITKA MOUNTAIN ASH
Sorbus sitchensis

Habitat: Woods and low elevation meadows.

Form: Shrubby tree — up to 18' (5.5m), alternate branches.

Leaves: Deciduous — pinnately divided, 7 to 11 leaflets (up to 2" long), toothed, rounded at ends, yellow to orange in fall, new buds have rusty hairs.

Flowers: June — white, small, 5 rounded petals (rose-like) in flat-topped clusters (corymb) at ends of branches.

Fruit: Mid-August to October — berry-<u>like</u>, red with bluish bloom.

Uses: Cooked, stewed — bitter when raw. Freezing and thawing removes bitter taste. A favorite food of Bohemian Waxwings.

Similar Species: Greene Mountain Ash (*Sorbus sambucifolia*) — winter buds are glossy, leaflets are slightly shorter, have 11 to 13 sharply toothed pointed segments. Fruit is orange to red and glossy. (See small map)

Sitka Mountain Ash

RED ELDERBERRY
RED-BERRIED ELDER
PACIFIC RED ELDER
Sambucus racemosa

Habitat: Woods and sub-alpine meadows.

Form: Shrub — up to 9' (3m), center of stems soft and pithy, surface rough, branches opposite, thick.

Leaves: Deciduous — pinnately divided, 5 to 7 toothed, pointed leaflets (2 to 5" long), coarse looking, opposite, strong objectionable aroma, winter buds large.

Flowers: Late May to early June — creamy white, very small, in conical panicle, pleasant aroma, stems pinkish.

Fruit: July to August — berry, red or orange, occasionally yellow, very small.

Uses: Cooked — jelly or juice.

Caution: POISONOUS WHEN RAW! — The seed contains a glycoside related to cyanide. This is destroyed by cooking.

Red Elderberry

ROSE HIP
PRICKLY ROSE
Rosa acicularis

Habitat: Open woods, clearings, meadows.

Form: Shrub — upright, with many prickles, reddish in winter.

Leaves: Deciduous — with stipules, alternate, compound, 5-parted, leaflets toothed, slightly hairy beneath, orange to maroon in fall.

Flowers: June — pink, 2 to 3", 5 rounded velvety petals, 5 long narrow sepals connected at base, many stamens.

Fruit: August and September — red, hip, oval to round, 3/4 to 1" (2 to 2.5cm), sweet, mild, soft when ripe, best after a frost. Very high in vitamin C.

Uses: Raw, cooked, dried, candied — jelly, jam, sauce, pie, juice, cake, tea (very good flavor). Must remove seeds (long hairs will irritate your intestines).

Other Species: Nootka Rose, *R. nutkatensis*, found in Southeast Alaska and coastal areas of Southcentral Alaska. Mature branches have only a few large prickles which are usually at the base of the leaves. Hybridization is common between the two species. Wood's Rose, *R. woodsii*, an introduced species, might be seen in Interior Alaska. The shrub, flowers, and hips are much smaller.

Prickly Rose

WESTERN CRAB APPLE
OREGON CRAB APPLE
Malus fusca

Habitat: Moist open woods of coastal areas.

Form: Small tree or shrub — up to 15' (5m), some stubby thorns, young twigs hairy, trunk rough and grooved.

Leaves: Deciduous — ovate to 3 lobed, toothed, mostly glabrous and dark green above, paler and hairy beneath.

Flowers: June to early July — pinkish-white, 5 rounded petals, in cluster at ends of branches.

Fruit: September — "pome" like a small apple, red to purplish, small, long oval.

Uses: Cooked — jelly, not abundant, rarely used.

Bark of Crab Apple

Immature fruit

Photo—Sally Karabelnikoff

Western Crab Apple

STRAWBERRY SPINACH
STRAWBERRY BLIGHT
Chenopodium capitatum

Habitat: Dry areas, waste places at low elevations.

Form: Plant — annual, usually branched, up to 24" (60cm), often sprawling.

Leaves: Glabrous, somewhat triangular with wavy edges, pointed, much like spinach leaves.

Flowers: Late June to July — greenish, very small in dense clusters on spikes above the leaves, often not noticed.

Fruit: August — fruit-like, dense clusters of reddish calyx, soft, juicy, sweet, mild.

Uses: Raw or cooked — jelly or syrup, low in pectin, not generally used.

Strawberry Spinach

WILD STRAWBERRY
Fragaria virginiana

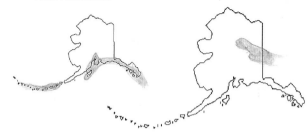

Habitat: Dry, open fields.

Form: Plant — perennial, up to 5" (12cm), with runners and new plantlets.

Leaves: Deciduous — 3-parted, toothed, on long slightly hairy stem, coarse veins.

Flowers: June — white, 5/8" (1.5cm), 5 rounded petals.

Fruit: July — accessory, fleshy red, small, sweet, mild.

Uses: Raw, cooked — jelly, jam, pie (excellent flavor).

Similar Species: Beach Strawberry, *Fragaria chiloensis*, (see small map). Commercial strawberries have escaped from cultivation in some areas and may cause some confusion when they grow in the wild. These plants are not as large and robust as those grown in gardens. However, usually their leaves are larger and more rounded than wild varieties.

Wild Strawberry

CLOUDBERRY
BAKED APPLE BERRY
SALMONBERRY
Rubus chamemorous

Habitat: Bogs, wet acidic woodlands and moist tundra.

Form: Plant — perennial, on creeping rootstock, up to 5" (12cm).

Leaves: Deciduous — 3 to 5 rounded, toothed lobes, coarse-veined, 1 to 3 per plant, red to yellow in fall.

Flowers: Late May to early June — white, 3/4 to 1" (2 to 2.5cm), 4 or 5 rounded petals (like apple blossoms), appearing with leaves, male and female on separate plants.

Fruit: Mid July to August — berry, aggregate fruit, soft, orange, seedy, tasty.

Uses: Raw, cooked — jelly, jam, (very good flavor). Produced in abundance only in wet areas.

Cloudberry

47

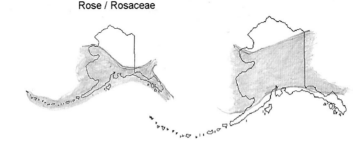

Rose / Rosaceae

NAGOONBERRY
WINEBERRY
Rubus arcticus

Habitat: Stream banks, fields, lake margins, tundra and alpine slopes.

Form: Plant — perennial, up to 5" (22cm) from creeping rootstock.

Leaves: Deciduous — on long stems, 3-toothed leaflets (like Strawberry), coarse veins.

Flowers: June — pink, 1 to 1-1/4" (2.5 to 3cm), 5 to 8 narrow petals rounded at ends, dark center.

Fruit: August — berry, aggregate fruit, red, soft, juicy, shiny, rounded, not found in abundance in most areas.

Uses: Raw, cooked — jelly, jam, pie. Calyx (hull) is sometimes hard to remove, but does not alter the taste if some is left on. (Excellent flavor).

Comments: 3 sub-species occur in Alaska with minor differences of petal width and leaf shape. The leaves of Ssp. stellatus are not divided to the base. (See small map).

Nagoonberry

TRAILING RASPBERRY
FIVE-LEAVED BRAMBLE
Rubus pedatus

Habitat: Moist, mossy woods from sea level to alpine.

Form: Plant — trailing rootstock.

Leaves: Evergreen — Shiny on long stems, 5-toothed leaflets.

Flowers: June to early July — white, 5 narrow spatulate widely-spaced petals.

Fruit: August and September — berry, aggregate fruit, red, soft, shiny, tasty, often found with only a few drupelets, ripen very late.

Uses: Raw, cooked — jelly, jam, pie. (Excellent flavor). Calyx (hulls) are hard to remove, but does not alter flavor if some are left on, difficult to prepare.

Trailing Raspberry

RASPBERRY
RED RASPBERRY
Rubus idaeus

Habitat: Clearings, dry meadows and edges of woods.

Form: Biennial canes — Canes develop the first year, bear fruit the second season and then die. Grows from horizontal rootstock, up to 5' (1.5m) tall, with prickles, yellowish-brown in winter.

Leaves: Deciduous — pinnately divided, 3 to 5 sharply pointed and toothed leaflets, coarse-veined and texture, dark above, whitish beneath.

Flowers: June — white, 5/8" (1.5cm), 5 widely spaced petals exposing 5 green sepals.

Fruit: July — berry, aggregate fruit, red, dull, with fine hairs, rounded, juicy, sweet to tart, sometimes crumbly.

Uses: Raw, cooked — jelly, jam, juice, desserts, liqueur (Excellent flavor).

Comments: Black Raspberry or Black Cap, *Rubus leucodermis*, may be seen in scattered locations in Southeastern Alaska. Flowers are white and fruit is purplish-black.

Raspberry

SALMONBERRY
Rubus spectabilis

Habitat: Stream banks, moist coastal meadows, alpine slopes and avalanche chutes.

Form: Biennial canes (see Raspberry) — up to 7-1/2' (2.5m), some prickles — taller and more robust than raspberries. Forms dense thickets.

Leaves: Deciduous — 3 to 5 leaflets, very coarse looking, (larger and darker than Raspberry), sharply pointed and deeply toothed, strongly veined.

Flowers: June — pink, 1" (2.5cm), 5 wide petals, pointed at end.

Fruit: Late July and August — berry, aggregate fruit, yellow, orange or red, round to oval, larger than Raspberry, sweet, juicy.

Uses: Raw, cooked — jelly, jam. (Very good flavor).

Salmonberry

THIMBLEBERRY
Rubus parviflora

Habitat: Along streams, forest edges and avalanche chutes.

Form: Shrub — up to 6' (2m), fuzzy brownish branches.

Leaves: Deciduous — 3 to 5-lobed, toothed, large.

Flowers: June — white, 1 to 1-1/2" (2.5 to 4cm), 5 rounded petals (rose-like).

Fruit: August and September — berry, aggregate fruit, dull, light red, hairy, often misshapen or flattened.

Uses: Raw, cooked — jelly, jam, pie, mediocre flavor, not widely used.

Thimbleberry

Photo-Clayton Antieau

Dwarf Blueberry
See page 60

BLUE OR BLACK BERRIES

Included in this section are berries that are blue to black or purplish-black when mature. Some are covered with a "bloom", a powdery substance that lightens their appearance. They may even appear white at first glance. Usually, this can be easily rubbed off. The presence of a bloom often helps in identification of some species.

All of these berries grow on shrubs or small trees and all berries are opaque, or dense – not translucent.

The most popular berries, by far, are Blueberries. Bog and Alpine Blueberries are the most widely used berries as they are tasty, plentiful, and easy to pick. Dwarf Blueberry has a sweeter flavor but is generally not found in abundance and the berries are difficult to pick because they hug the ground. Tall Blueberries, or Huckleberries, are more firm (Don't crush as easily in your berry bucket or freezer), but have larger seeds. They are easy to pick but are not preferred by most people.

At the back of this section we have listed the Junipers. These actually have blue cones, that somewhat resemble a berry when they are mature.

BOG BLUEBERRY
BOG BILBERRY
Vaccinium uliginosum

Habitat: Bogs, woodlands, wet and dry tundra up into alpine.

Form: Shrub — upright up to 2' (60cm), prostrate in alpine areas, thin branches.

Leaves: Deciduous — oval, 5/8 to 3/4" (14 to 18mm), alternate, turning orange, red or purplish in the fall.

Flowers: June — pinkish, small, urn-shaped, opening with the leaves.

Fruit: Mid July to early September — berry, dark blue with a lighter bloom, round to oval, juicy, tart, acidic.

Uses: Raw, cooked, dried — jelly, jam, pies, desserts, breads, syrup, tea. (Excellent flavor).

Comments: There are two sub-species in Alaska, *Vaccinium alpinum and Vaccinium microphyllum.*

Similar Species: Dwarf Blueberry, *Vaccinium caespitosum,* (see small map) is a low alpine shrub having very small leaves with fine teeth and round, light blue, quite sweet berries. See page 58.

Bog Blueberry

61

EARLY BLUEBERRY
BLUE HUCKLEBERRY
Vaccinium ovalifolium

Habitat: Moist, coastal forests, meadows.

Form: Shrub — upright to 3' (1m), branches reddish in spring.

Leaves: Deciduous — ovate, thin, 3/4 to 1-1/2" (18 to 37cm) long.

Flowers: May to early June — pink, urn-shaped, small, usually longer that broad, usually opening before the leaves.

Fruit: Late July and August — berry on a curved stem, round, blue to purplish with a bluish bloom, juicy, somewhat seedy, firm.

Uses: Raw, cooked — jelly, jam, syrup, pie, desserts, pancakes (good flavor).

Early Blueberry

ALASKA BLUEBERRY
Vaccinium alaskensis

Habitat: Moist coastal forests.

Form: Shrub — upright to 6' (2m), new twigs yellowish-green.

Leaves: Deciduous — ovate, usually shallowly toothed and having some glands.

Flowers: Late May to June — small, greenish-white to rosy or coppery, urn-shaped, usually broader than long, appearing after the leaves.

Fruit: Late July and August — berry, black, shiny, usually on straight stem, flattened end. Easily confused with Early Blueberry.

Uses: Raw, cooked — jelly, jam syrup, pie, desserts, pancakes (good flavor).

64

Alaska Blueberry

PACIFIC SERVICEBERRY
JUNEBERRY
SASKATOON BERRY
Amelanchier florida

Habitat: Dry, exposed areas from sea level up into sub-alpine.

Form: Shrub — 2 to 12' (0.6 to 4m).

Leaves: Deciduous — alternate, 1-1/2 to 2-1/2" (4 to 6cm), toothed, broadly oval, longer than broad, creased at mid-vein.

Flowers: Late May to early June — white, 5 narrow petals, terminal cluster, sepals slightly hairy.

Fruit: Late August and September — berry, dark blue, round with pronounced blossom-end, sweet, not very juicy, large triangular seeds. Often mistaken for Blueberries.

Uses: Raw, cooked, dried — jam, jelly, pies, dried (like raisins). Good flavor.

Other Species: Alder-leaf Serviceberry, *Amelanchier alnifolia*, (See small map). Leaves are about as broad as long and sepals are densely hairy.

Service Berry

SALAL
LAUGHING BERRIES
Gaultheria shallon

Habitat: Coastal woods.

Form: Shrub — 2 to 3 feet (up to 1m), often sprawling; appearing to be less than a foot tall. Twigs with reddish hairs.

Leaves: Evergreen — thick, leathery, ovate, 2 to 4" (5 to 10cm), with small sharp teeth, shiny above with some reddish hairs and obvious raised veins.

Flowers: June — pink, urn-shaped, on long curved racemes, hanging down. Calyx is brown, hairy and sticky.

Fruit: August — purplish black, juicy.

Uses: Raw, cooked or dried — flavor varies which may explain why they are not often used. Healthy shrubs in moist areas produce very tasty berries that are eaten raw or made into jam.. Dry areas may produce tasteless or distasteful berries.

Comments: The common name of Laughing Berries has an interesting origin. The juice from the purplish-black berries stains the eater's mouth and lips, giving the appearance of a broad smile.

68

Photo—Forrest Baldwin   Salal   Ripe berries   Photo—Frank Lang

ALPINE BEARBERRY
PTARMIGAN BERRY
Arctostaphylos alpina

Habitat: Dry tundra.

Form: Shrub — dwarf, forms mats.

Leaves: Deciduous — spatulate, coarse-veined, wavy margins, some hairs on edges, scarlet to maroon in the fall.

Flowers: May to early June — creamy-white, urn-shaped, usually before the leaves emerge.

Fruit: August and September — berry, black, shiny, opaque, round, juicy, beneath leaves.

Uses: Raw or cooked — fair flavor, not widely used, mostly as an extender.

Alpine Bearberry

CROWBERRY
MOSSBERRY
BLACK BERRY
Empetrum nigrum

Habitat: Woods, bogs, wet and dry tundra (especially north-facing slopes).

Form: Shrub — somewhat decumbent, up to 8" (20cm).

Leaves: Evergreen — dark green, small, narrow, needle-like, wine or maroon color in early spring.

Flowers: May to early June — maroon, inconspicuous, in parts of three.

Fruit: August and September — berry, black, shiny, opaque, round, sweet, juicy, seedy.

Uses: Raw or cooked — jelly, pies (good flavor). Has very low pectin content.

Crowberry

NORTHERN BLACK CURRANT
Ribes hudsonianum

Habitat: Moist open woods and clearings.

Form: Shrub — upright, up to 3' (1m), winter buds have resin glands, young twigs are light brown.

Leaves: Deciduous — 3 to 5-lobed (maple-like), toothed, yellow in fall, strong odor.

Flowers: Late May to June — white, 5 small flaring sepals in short erect racemes.

Fruit: July and August — berry, black, smooth, opaque, in short slightly drooping racemes. Very sour, strong flavor, somewhat bitter.

Uses: Raw or cooked — jelly, jam, not widely used. Substitute 1/2 cup of Blueberries in a Blueberry pie recipe with 1/2 cup of Northern Black Currant for an excellent pie.

Northern Black Currant

TRAILING BLACK CURRANT
TRAILING BLUE CURRANT
Ribes laxiflorum

Habitat: Woods, rocky outcroppings up into alpine.

Form: Shrub — up to 3' (1m), sprawling and sometimes vine-like with long branches, winter buds reddish.

Leaves: Deciduous — 3 to 5-lobed (maple-like), more deeply lobed and toothed than Northern Black Currant, turning orange to red in fall, strong odor.

Flowers: Late May to early June — usually rosy, sometimes greenish, hanging in long racemes under branches.

Fruit: Late July through August — berry, black with bluish bloom, hairy, sweet, juicy, strong musty flavor.

Uses: Raw, cooked or dried — jelly, jam, liqueur, good flavor but not widely used.

Trailing Black Currant

Photo—Forrest Baldwin

STINK CURRANT
Ribes bracteosum

Habitat: Moist coastal areas at low elevations.

Form: Shrub — usually upright, to 7' (2m).

Leaves: Deciduous — large, 3 to 8" (up to 20cm), 5 to 7-lobed, deeply lobed and toothed, underside dotted with resin glands, strong smelling.

Flowers: June — greenish to rusty in very long racemes.

Fruit: August — berry, black or with bluish bloom, hairy, sour.

Uses: Raw or cooked — Not generally used. Not usually tasty.

Stink Currant

SWAMP GOOSEBERRY
BRISTLY BLACK CURRANT
Ribes lacustre

Habitat: Moist woods.

Form: Shrub — 2 to 3' (up to 1m) with thorns, usually upright.

Leaves: Deciduous — small, 1 to 1-1/2" (2.5 to 3.5cm), 3 to 5-lobed (maple like).

Flowers: Late May to mid June — 5 rosy-green sepals, small, in short racemes that hang down.

Fruit: August — berry, black, shiny, hairy, juicy, very sour.

Uses: Raw or cooked — jellies, pie, syrup (good flavor).

Similar species: Northern Gooseberry, *Ribes oxycanthoides*, may be found in Southcentral Alaska. (See small map) Berries are single or in pairs and purplish-black.

Photo-Forrest Baldwin
Swamp Gooseberry

Northern Gooseberry (unripe)

81

BLACK HAWTHORN
Crataegus douglasii

Habitat: Moist open forests and coastal bluffs.

Form: Small tree or shrubby tree — up to 12' (3.5m).

Leaves: Deciduous — shiny, dark green above, ovate, sometimes lobed, broader above the middle, finely toothed.

Flowers: June — white, small, rose-like, in clusters, 5 petals and 10 stamens.

Fruit: August and September — berry-like (like very small apples), purple to black, dry.

Uses: Cooked — Not generally used.

Black Hawthorn

Photos—Rhoda M. Love

COMMON JUNIPER
MOUNTAIN JUNIPER
Juniperus communis

Habitat: Dry, rocky outcroppings, dry woods and alpine ridges.

Form: Shrub — sprawling, usually less than 2' (60cm), shredding bark.

Leaves: Evergreen — narrow, needle-like, stiff, prickly, dark green above, whitish below, very prickly.

Flowers: Seldom noticed, very small, male and female on separate plants.

Fruit: Cone (not a berry) — requires 2 years to develop, blue when mature, hard, bitter.

Uses: Ground and used as a seasoning for meats, and flavoring in the production of gin.

Similar Species: Horizontal Juniper or Creeping Savin, *Juniperus horizontalis,* is common in very dry sunny areas of Southcentral, Southeastern, Eastern and Interior Alaska. This is a trailing evergreen shrub. Leaves are small, scale-like, lying close to the stems — not prickly. Cones are blue when mature.

Common
Juniper

Horizontal
Juniper

SILVERBERRY
Elaegnus commutata

Habitat: Sandy or gravelly riverbeds, dry glacial banks and sunny dry low elevation slopes.

Form: Shrub — up to 10' (3m), alternate branches, main trunk blackish with age. Young twigs silvery with brownish scales.

Leaves: Deciduous — alternate, silver with rough rusty scales (like sandpaper), ovate, 2 to 3" (10 to 12cm).

Flowers: June — yellow, salverform, 4 sepals, silver on underside, aromatic (like Gardenia). Close to main stem.

Fruit: August — berry, oval, silver with scales, dry, mealy, tasteless.

Uses: Not generally used now. Traditionally used by some ethnic groups.

This is the only berry in Alaska that falls in this category, and it is rarely used for food. The berry of Silverberry is the same color as the leaves of the shrub, so it is frequently overlooked by both humans and animals. The shrubs also strongly resemble willows which do not produce berries.

Silverberry

Photo—Ruth Munson

False Solomon's Seal (see page 92)

POISONOUS AND INEDIBLE

This is probably the most important section of this book. Recognizing poisonous berries is of utmost importance. Baneberry (which means "Poison Berry") is the most dangerous. Some of the others may merely be considered inedible due to stomach upsets, allergy-like symptoms, or perhaps just being bitter or distasteful. You might even find some of these listed as edible in other publications.

Do not rely on what birds and animals eat as their digestive systems are not the same as that of humans. They sometimes eat things that taste terrible to humans or perhaps are even poisonous. Neither should you rely on a taste test; although many poisonous substances taste bad, some do not.

Caution and close attention to detail is highly recommended. Stop, look, look again and again.

If anyone you are with does happen to eat a questionable berry or becomes ill after eating some berries call the local Poison Control Center. The number for your local area is usually located in the front of your telephone directory. If possible, it is recommended that you gather a sample of the berry and plant or at least be prepared to provide a very good description of both.

**BANEBERRY
SNAKE BERRY
DOLL'S EYES**
Actaea rubra

Habitat: Woods and dry hillsides.

Form: Plant — perennial, dark-colored in spring.

Leaves: Deciduous — all from one stalk, 1 nearly basal, 1 or 2 smaller above it. Large, 3 to 5-parted, finely dissected and toothed, narrow pointed leaflets, appearance and width of leaf segments change radically with the season, narrow and crinkled in the spring — very broad in summer.

Flowers: May to early June — white, small, delicate in rounded clusters above the leaves.

Fruit: July and August — berry, red or white, opaque, shiny with a black dot. On an elongated stem, each berry also has its own stem.

Comments: There are two subspecies in Alaska; Ssp. rubra being interior, and Ssp. arguta being coastal. (See small map).

POISONOUS: A few berries, if eaten, can kill a small child!

Baneberry

STAR-FLOWERED SOLOMON'S SEAL
FALSE SOLOMON'S SEAL
Smilacina stellata

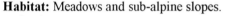

Habitat: Meadows and sub-alpine slopes.

Form: Plant — perennial from creeping rootstock, 6 to 24" (15 to 60cm).

Leaves: Alternate on an unbranched, stiff, slightly arching stem, stalkless, long, broad at base tapering to a point, often angled upwards.

Flowers: Late May to early June — white, small, 6 tepals, star-like in terminal raceme.

Fruit: August and September — dark red (sometimes streaked), 3-parted, hard, not palatable.

Comments: Considered inedible, possibly poisonous.

Similar species: *Smilacina racemosa*, also called False Solomon's Seal, is found in woodlands in the extreme southern part of Southeast Alaska. Leaves are broader (more wide spread), flowers are in a loose fluffy raceme. Berries are similar and considered inedible. See page 90.

Star-flowered Solomon's Seal

DEVIL'S CLUB
Echinopanax horridum

Habitat: Wet coastal forest.

Form: Shrub — up to 12' (3m), from creeping
rootstock, often unbranched, covered with spines.

Leaves: Deciduous — on long petioles, very large (maple-shaped), up to 15" (36cm), 5
to 9-lobed with sharp points and teeth, palmately veined, spines on veins, leaf buds
large.

Flowers: June — small, green, in small umbels on an erect raceme above the leaves.

Fruit: August — berry, red, small oval, with 2 spines, soft.

Comments: Not recommended, considered to be toxic.

Devil's Club

Photo—Michael MacDonald

WILD CALLA
Calla palustris

Habitat: In shallow water along the edges of lakes and slow-moving streams.

Form: Plant — perennial, from creeping rootstock, thick stems.

Leaves: Heart-shaped on thick stem, thick, shiny, palmately veined.

Flowers: June and July — greenish, very small on a stout dense spike atop a large white heart-shaped spathe.

Fruit: August — berry, red, soft.

Comments: <u>POISONOUS</u>, entire plant contains poisonous acids and saponin-like substances.

Skunk Cabbage

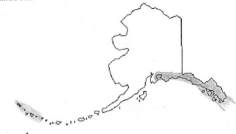

FALSE LILY OF THE VALLEY
DEERBERRY
WILD LILY OF THE VALLEY
Maianthemum dilatatum

Habitat: Moist woods and meadows, and wet tundra.

Form: Plant — perennial from creeping rootstock, 8 to 12" (20 to 30cm).

Leaves: Deciduous — heart-shaped, one to three per stem, alternate, shiny, with palmate veins.

Flowers: June — creamy white, small, feathery, many on a long stalk above leaves, scented.

Fruit: July or August — berry, mottled brown and red.

Comments: Considered inedible, contains cardiac glycoside that is active on the heart.

False Lily of the Valley

BLACK TWINBERRY
BEARBERRY HONEYSUCKLE
HONEYSUCKLE
Lonicera involucrata

Habitat: Moist woods in a few scattered spots in Southeastern Alaska.

Form: Shrub — up to 6' (2m), opposite branches, 4 angled branches.

Leaves: Deciduous — lance-shaped, up to 5" (12cm), underside with coarse veins.

Flowers: June — yellow, tubular, in pairs, on top of large greenish bract.

Fruit: August — black, soft, round, bracts turn red and become reflexed.

Comments: Not tasty, considered inedible.

Black Twinberry

SINGLE-FLOWERED CLINTONIA
QUEENS CUP
BLUE BEAD
Clintonia uniflora

Habitat: Moist forests, low elevations.

Form: Plant — perennial from creeping rootstock.

Leaves: Oblanceolate, 3 to 6" (7 to 15cm), 3 per plant, fleshy, hairy edges.

Flowers: June — white, about 1" (2.5cm), 6 tepals.

Fruit: August — berry, metallic blue, soft.

Comments: Not palatable, considered inedible.

Single-flowered Clintonia

Photo—Dave DeMoss

SNOWBERRY
WAXBERRY
Symphoricarpus albus

Habitat: Woods at low elevations.

Form: Shrub — up to 4' (1.3m), opposite branches, fine twigs.

Leaves: Deciduous — oval to ovate, opposite, 3/4 to 2" (2 to 5cm), with irregular teeth, dark green above, whitish and hairy beneath.

Flowers: June — pinkish white, small, bell-shaped.

Fruit: August — berry (drupe), white, round, soft, opaque, with 2 seeds.

Comments: Not considered edible.

Snowberry

RED-TWIG DOGWOOD
RED-OSIER DOGWOOD

Cornus stolonifera

Habitat: Woods, margins of lakes at low elevations.

Form: Shrub — from creeping rootstock, 5 to 15' (1.5 to 4.5m), opposite branches, reddish (especially in winter) with white dots.

Leaves: Deciduous — elliptical to ovate with arcuate veins, dark above, lighter and somewhat hairy beneath, turning yellow to copper color in fall.

Flowers: June — white to greenish, four tiny green sepals atop small white bracts, in flat-topped clusters.

Fruit: August — berry, white, soft, dark spot at end, bitter.

Comments: Not generally used — considered inedible by most, but was once used by some ethnic groups.

Red-Twig Dogwood

PICTORIAL GLOSSARY

Branch and leaf arrangements

Alternate branches and leaves

Opposite branches and leaves

Plant Types

Growing from rhizomes

Basal with runners

with stolons

Compound Leaves

Simple leaves

LEAF SHAPES AND VEINS

Ovate
Entire
margins

Oval
with
pinnate
veins

Elliptical
with
arcuate
veins

Spatulate
nearly
parallel
veins

Heart-shaped
with
palmate
veins

Toothed margins

Entire margin

Lobed

Maple-like

GLOSSARY

Acute: Sharply pointed.

Aggregate: In reference to fruit type. Separate carpels grouped together to form a fruit. Examples: Raspberry and cloudberry.

Alpine: Growing above timberline.

Alternate: Leaf or branch arrangement on stem, not opposite each other.

Annual: A plant growing from seed, blooming, setting seed and then dying all in one growing season.

Arching: usually referring to a branch or stem, curved, bending as half of an arch.

Arcuate: Usually referring to veins of a leaf that are bowed or follow the curve of the leaf and appear to be nearly parallel.

Berry: A soft, fleshy, multi-seeded fruit.

Biennial: Of two seasons duration from seed to maturity and death.

Bloom: A whitish coating on a fruit that, usually, can be rubbed off. Often waxy looking.

Blossom end: The end of a berry away from the stem, often having remains of the sepals still attached.

Bog: A low, very wet area. Soil is often acidic, and standing water is common.

Bract: A reduced or modified leaf, usually below a flower, often petal-like.

Calyx: The outermost circle of the floral parts. The external portion, usually green. The group of sepals.

Cane: A somewhat woody, usually unbranched, stem, usually a biennial.

Clasping: Partially surrounding the stem (usually referring to a leaf petiole).

Deciduous: Not persistent, said of leaves falling in autumn or of floral parts falling after flowering.

Decumbent: Sprawling on the ground with upturned ends.

Drupe: A fleshy fruit having one seed inside a tough shell.

Drupelets: Small drupes, usually in a group as a Raspberry.

Edible: Allright to eat — not poisonous.

Ethnic: Pertaining to a person, or group of people, who are native to, or have traditionally lived in, an area.

Evergreen: Remaining green all year.

Extender: A use for non-flavorful berries, especially in poor berry production years.

Glabrous: Having a smooth, even surface without hairs.

Hip: The fruit of a Rose — much like a dry apple

Hull: The calyx of a berry — the part that holds an aggregate berry together.

Hummock: A rounded rise of vegetation, usually in a wet area; such as, a bog or marsh.

Hybrid: A cross between 2 species or subspecies, usually of the same genus.

Inedible: Opposite of edible — not good to eat, perhaps because of an unpleasant taste or poisonous activity.

Insipid: Without flavor, not tasty.

Lance-shaped: Long and narrow, tapering toward both ends.

Margin: Edge.

Mediocre: Not choice, not flavorful.

Mottled: Blotchy or variegated.

Oblanceolate: Similar to lanceolate, but broader at base.

Opaque: Not letting light through, not transparent nor translucent.

Ovate: Egg-shaped, with a point.

Palatable: Pleasing to the palate, tasty.

Palmate: Lobed, divided or ribbed so as to resemble the outstretched fingers of a hand.

Parallel: Lines that run equidistant beside each other without crossing.

Parasitic: An organism obtaining food and/or shelter at the expense of another.

Perennial: Living for more than two years; and, usually, flowering each year after the first.

Petal: Usually the colorful part of the corolla, the row of floral parts within the sepals.

Petiole: The stalk that attaches the leaf to the stem.

Pinnate: Describing a compound leaf in which the leaflets are arranged in two rows, one on each side of the midrib.

Pithy: Soft and full of holes.

Plantlets: Small plants that develop while still attached to the adult plant.

Prickle: A short woody pointed outgrowth from the outer layer of a plant.

Raceme: An inflorescence in which the flowers are formed on individual pedicels attached to the main stem.

Reflexed: Bent abruptly downward or backward.

Resin glands: Openings in a stem or leaf surface allowing a thick secretion to escape.

Runner: A stolon or trailing stem.

Scale: A small projection on the surface of a leaf or stem causing it to feel rough like sandpaper.

Sepal: One of the outer group of floral parts. Usually green .

Shredding: Peeling off in small narrow strips.

Shrub: A woody perennial, smaller than a tree, usually with several basal stems.

Spathe: A large bract, or pair of bracts, often petal-like, enclosing a flower cluster or spadix.

Spatulate: Describing structures that have a broad end and a long narrow base, such as the leaves of the daisy.

Spines: Long pointed ends of thorns.

Stipule: A part of the leaf where it attaches to the stem (especially common in the rose and pea families). A broadening on either side of the stem that could be mistaken for a bract.

Sub-alpine: Below treeline.

Sub-shrub: A very low shrub.

Tepals: One name for petals and sepals that look alike (mostly in the Lily family).

Toothed: Having teeth or jagged or notched edges.

Translucent: Allowing some light through without being able to see inside.

Tree: A plant with woody parts above the ground — usually having one trunk.

Treeline: The elevation above which trees no longer grow.

Tundra: Treeless land, often times damp, but can also be dry and rocky.

Variegated: Having more than one color, usually blending in a pleasing way.

PLANT FAMILIES

Plants are divided into families by differences of reproductive parts; such as, number and placement of stamens, how the ovaries are divided, placement of seeds within the ovaries, manner of seed dispersement, etc. However, these characteristics are not always available or easily noticed, so I have listed below some other, more obvious, characteristics to look for. In botany, there are a lot of "usually"s, some families vary greatly, and many oddities do occur. Included here are the plant families in this book.

(1) Buttercup or Crowfoot / Ranunculaceae — herbaceous plants with 1 to many common flowers usually having 5 sepals, 5 petals, many stamens (tight cushion effect). This is a very variable family having unusual numbers of petals or sepals; and, sometimes completely lacking petals. Leaves are frequently divided into a lobed or dissected crowfoot (birdfoot) pattern, have long stems, and are predominantly basal. Many members of this family are poisonous. Baneberry is the only member of this family represented in this book.

(2) Calla / Araceae — herbaceous plants having large, simple leaves and a large floral type leaf below a spike of inconspicuous flowers having no petals and 6 stamens. The fruit is a berry, frequently poisonous.

(3) Crowberry / Empetraceae — evergreen shrubs having inconspicuous flowers of 3 bracts, 3 to 6 sepals (sometimes in whorls and confused as petals) and 2 to 4 stamens. The ovary is divided into many parts and produces a round, black berry. Leaves are simple and heath-like.

(4) Cypress / Cupressaceae — evergreen trees or shrubs with needle-like or scale-like leaves that may be pressed close to the branches. Producing woody cones, or in the Junipers somewhat fleshy, berry-like cones.

(5) Dogwood / Cornaceae---shrubs or sub-shrubs having clusters of very small flowers consisting of 4 showy bracts, 4 sepals, 4 petals, 4 stamens, and 1 ovary which produces a soft berry. The 4 large, showy bracts are often confused with petals. The simple leaves have arcuate veins.

(6) Ginseng / Araliaceae — shrub with umbels of very small flowers on a raceme atop a spiny upright branch. Flowers have 5 sepals, 5 petals, 5 stamens and 1 ovary that forms a small berry. Large maple-like leaves with palmate veins.

(7) Gooseberry / Grossulariaceae — shrubs sometimes with thorns, having small flowers with 4 or 5 sepals, 5 very small inconspicuous petals, 5 stamens, and a 2-parted ovary in the form of a berry. Leaves are usually lobed with teeth.

(8) Goosefoot / Chenopodiaceae — herbaceous plants with spikes of small, mostly green, inconspicuous clusters of flowers with 5 sepals, no petals, 5 stamens, and 1 ovary. These are mostly weedy plants usually with opposite leaves.

(9) Heath / Ericaceae---mostly shrubs, frequently evergreen, with bell or urn or cup-shaped flowers. Flowers have 4 or 5 (sometimes united) sepals, 4 or 5 (usually united) petals, 4 or 5 stamens, and 1 ovary. The leaves are entire, simple and, usually narrow. Usually found in acidic soil.

(10) Honeysuckle / Caprifoliaceae---shrubs with tubular or salverform flowers having 5 sepals, 5 united petals, 5 stamens, and 1 ovary. Leaves are toothed and opposite.

(11) Lily / Liliaceae---frequently bulbous plants usually with a stout flower stalk. Flowers have 6 tepals (3 sepals, 3 petals), 6 stamens, and a 3-parted ovary. Flowers are in a raceme or umbel. Leaves have parallel veins and, frequently, clasp the stem.

(12) Oleaster / Elaegnaceae---shrubs having salverform flowers with 4 sepals, no petals, 4 stamens, and 1 ovary in the form of a berry. Leaves are simple, entire, have scales, and can be opposite or alternate on the stem..

(13) Rose / Rosaceae---plants or shrubs with flowers having 5 sepals, usually 5 petals, many stamens, and one to many fruits. Leaves are varied, but most are pinnately divided and have stipules.

(14) Sandlewood / Santalaceae---plants, sometimes parasitic, with 3 to 5 sepals and no petals. Leaves are usually simple and alternate, often producing a berry.

BIBLIOGRAPHY

Heller, Christine. 1953. **Wild Edible and Poisonous Plants of Alaska**. Univ. Alaska
Ext. Bull. F-40. 87 pp.

Hultén, Eric. 1968. **Flora of Alaska and Neighboring Territories**. Stanford University
Press. 1008 pp.

Pojar, Jim and MacKinnon, Andy. 1994. **Plants of the Pacific Northwest Coast.** Lone
Pine Publishing, #180, 16149 Redmond Way, Redmond, Washington. 527 pp.

Smith, James P., Jr. 1977. **Vascular Plant Families**. Mad River Press, Eureka,
California. 321 pp.

Viereck, Leslie A., and Little, Elbert L., Jr. 1972. **Alaska Trees and Shrubs**. Forest
Service, U. S. Dept. of Agriculture. Washington, D. C. 266 pp.

Welsh, Stanley L. 1974. **Anderson's Flora of Alaska and Adjacent Parts of Canada**.
Brigham Young University Press, Provo, Utah. 724 pp.

INDEX

Ordering information on publications by Alaskakrafts, Inc.

Each of these flower books is arranged by flower color and have color bars on the edges of the pages. All are approximately 6 x 9 vertical format, except "Wild Berries' which is 4-1/4 x 5-3/4" in horizontal format.

Field Guide to Alaskan Wildflowers $15.95
136 pages - 248 color photos - A guide to the most common flowers of Alaska, in areas accessible by road. A very useful book for the novice and residents of Southeastern and Southcentral Alaska.

Wildflowers along the Alaska Highway $19.95
223 pages - 497 color photos - A guide to ALL the flowers we saw from Dawson Creek, B.C., Canada to Fairbanks, Alaska. An indispensable guide for persons traveling the "ALCAN" and for residents of Interior Alaska.

Wildflowers of Denali National Park (and Interior Alaska) $14.95
176 pages - 410 color photos - A guide to most of the plants of Denali National Park. An indispensable guide for tourists or hikers. This is also a very useful addition to the Field Guide as it includes many different alpine plants and is a great guide for Interior Alaska.

Alaska's Wild Berries $9.95

ORDER FORM

Name: _____

Address: _____

City: _____ State: _____ Zip: _____

Telephone: (in case of questions about your order)_____

How Many	Order Code	Price each	Extended Price
	Alaska's Wild Berries	$9.95	
	Field Guide to Alaskan Wildflowers	$15.95	
	Wildflowers along the Alaska Highway	$19.95	
	Wildflowers of Denali National Park	$14.95	
		Plus shipping	
		TOTAL	

Book Rate: $2.00 for first book and 75 cents for each additional book (Surface shipping may take three to four weeks) Air Mail: $3.00 for first book and $1.50 for each additional book.

Payment: ˇ Check   ˇ Money Order   ˇ Visa   ˇ Mastercard   ˇ AMEX

Card# _____Exp._____

Signature:_____

Make checks payable to: Alaskakrafts, Inc., 7446 East 20th, Anchorage, AK 99504-3429
Tel-(907) 333-8212,   Fax-(907) 333-4989,   E-mail-akkrafts@alaskakrafts.com